Contents

Words printed in **bold** in the main text are explained in the glossary on page 45.

What is the atmosphere?

It is all around us but we cannot see it. What is it? Air, or as it is also called, the **atmosphere**.

▼ *A hang glider flies through the air – without air, it would drop like a stone.*

Atmosphere

John Baines

Our Green World

Acid Rain
Atmosphere
Deserts
Farming
Oceans
Polar Regions
Rainforests
Recycling
Wildlife

Cover: Clouds are formed from tiny drops of water in the atmosphere.
These are clouds seen from Maui in Hawaii, USA.

Book editor: Sue Hadden
Series editor: Philippa Smith
Series designer: Malcolm Walker

First published in 1991 by
Wayland (Publishers) Ltd
61 Western Road, Hove
East Sussex BN3 1JD, England

British Library Cataloguing in Publication Data
Baines, John
    Atmosphere.—(Our green world)
    I. Title II. Series
    551.5

HARDBACK ISBN 0–7502–0136–3

PAPERBACK ISBN 0–7502–0587–3

Typeset by Kudos Editorial and Design Services, Sussex, England
Printed in Italy by G. Canale & C.S.p.A., Turin

◀ *Flashes of lightning jump through the air during a thunderstorm.*

The Earth is surrounded by the atmosphere. It is a mixture of **gases**, moisture and very tiny pieces of dust and dirt. The two main gases are oxygen and nitrogen. All animals need to breathe oxygen to stay alive. There are small amounts of other gases as well, such as **carbon dioxide**. Plants use carbon dioxide as they grow.

Why is air important?

Without air, there would be no water, plants, animals or people.

▼ *These flowers could not grow without air.*

This is a photograph of the Earth taken from a spacecraft. The white patterns are clouds in the atmosphere. ▶

The Earth's atmosphere is made up of different layers. ▼

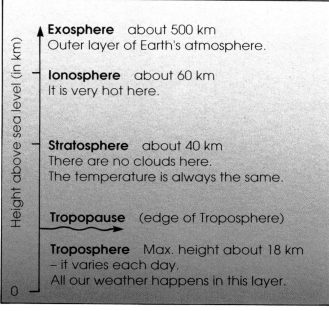

Height above sea level (in km)

Exosphere   about 500 km
Outer layer of Earth's atmosphere.

Ionosphere   about 60 km
It is very hot here.

Stratosphere   about 40 km
There are no clouds here.
The temperature is always the same.

Tropopause   (edge of Troposphere)

Troposphere   Max. height about 18 km
– it varies each day.
All our weather happens in this layer.

We need to breathe air to stay alive. Stop for a moment and count how many times you breathe in and out in one minute.

All plants and animals need air to stay alive.

The air **absorbs** moisture from the sea, lakes, rivers and the soil. Plants and animals also give off some moisture. The wind carries the moisture in clouds, from one place to another. Moisture falls back to the Earth as rain, mist or snow. This movement of water from the Earth's surface, into the air and back again, is called the water cycle.

The wind is air moving from one place to another. The world's winds follow regular patterns, as this map shows. ▶

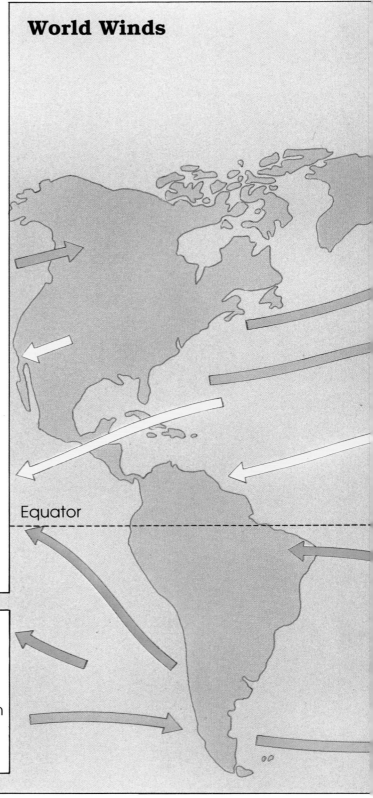

World Winds

Equator

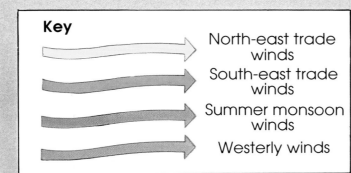

Key

North-east trade winds

South-east trade winds

Summer monsoon winds

Westerly winds

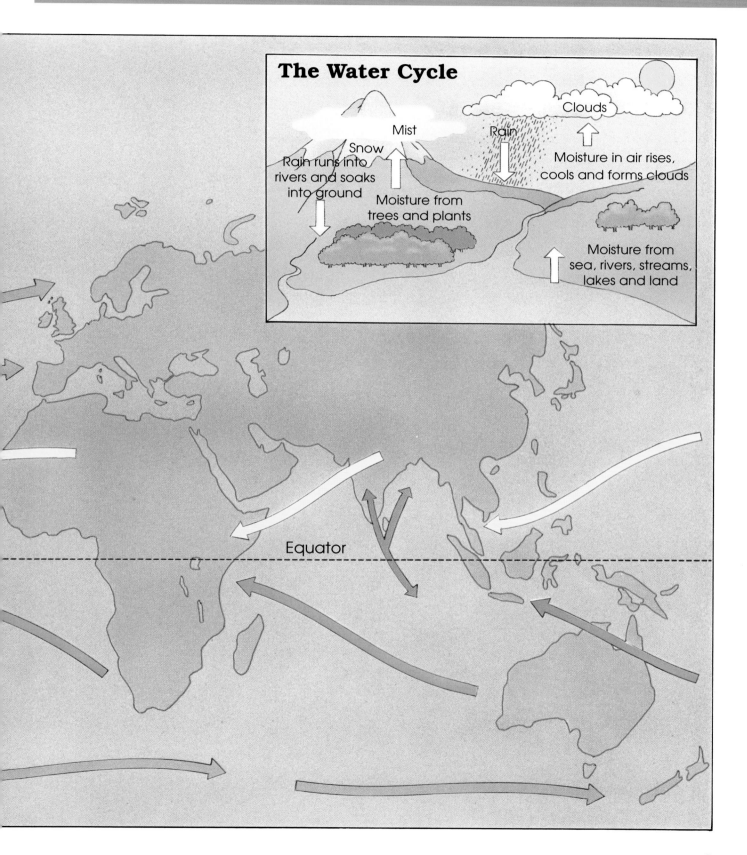

The Water Cycle

Clouds

Mist

Rain

Snow

Rain runs into rivers and soaks into ground

Moisture in air rises, cools and forms clouds

Moisture from trees and plants

Moisture from sea, rivers, streams, lakes and land

Equator

Why is the air dirty?

We burn **fossil fuels** (coal, oil and gas) in **power stations** to create energy for our homes and factories.

◀ *Smoke from chimneys, and steam from cooling towers, rises into the* ▼ *atmosphere.*

▲ *The Thames Barrier was built to protect London from being flooded by the sea.*

Coal, oil and gas also produce smoke and **fumes**. Some of this goes into the air. When the air is very dirty we say it is polluted. Air **pollution** is making the Earth warmer. Over the next 50 years, the temperature will probably rise by another 1.5°C.

As the Earth warms up, some of the ice around the North and South poles will melt. This will raise the level of the oceans and cause floods. In 50 years' time, the sea level could be one metre higher than today.

HURRICANE GILBERT
SEPT. 15 1988
9 AM EDT

▲ *This is a satellite photograph of Hurricane Gilbert. It caused great damage in the Caribbean in 1988.*

Many of the Earth's big cities and best farming areas are near the sea. High banks may need to be built to prevent flooding.

◀ *A flood in the city of Calcutta in India. If the oceans grow, floods will be more common.*

◄ *In 1987 and 1989 Britain had extremely strong winds.*

In Jamaica the winds of Hurricane Gilbert threw this aeroplane around. ▼

▲ *An American farmer looks at his small maize plants. There was too little rain for them to grow properly.*

The weather is changing. Storms happen more often and cause damage. The rainfall in parts of Africa, Australia and the USA is getting less. Crops need rain to grow and many people are not getting enough to eat.

The Earth and the air above it are warmed by the Sun. The heat then escapes slowly into space. Some gases in the air, such as carbon dioxide, do not let all the heat escape. These are called 'greenhouse gases'. Like glass in a greenhouse, they let in the Sun's heat and trap it.

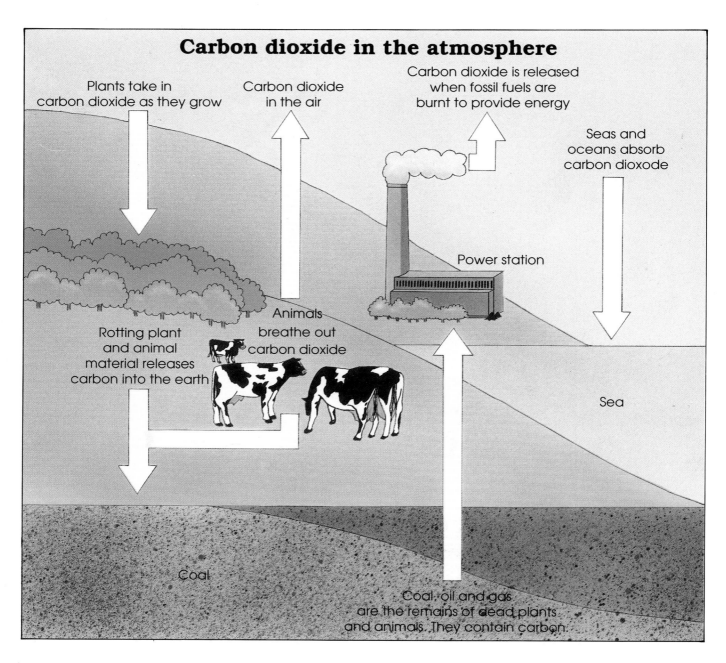

Carbon dioxide in the atmosphere

Plants take in carbon dioxide as they grow

Carbon dioxide in the air

Carbon dioxide is released when fossil fuels are burnt to provide energy

Seas and oceans absorb carbon dioxode

Power station

Rotting plant and animal material releases carbon into the earth

Animals breathe out carbon dioxide

Sea

Coal

Coal, oil and gas are the remains of dead plants and animals. They contain carbon.

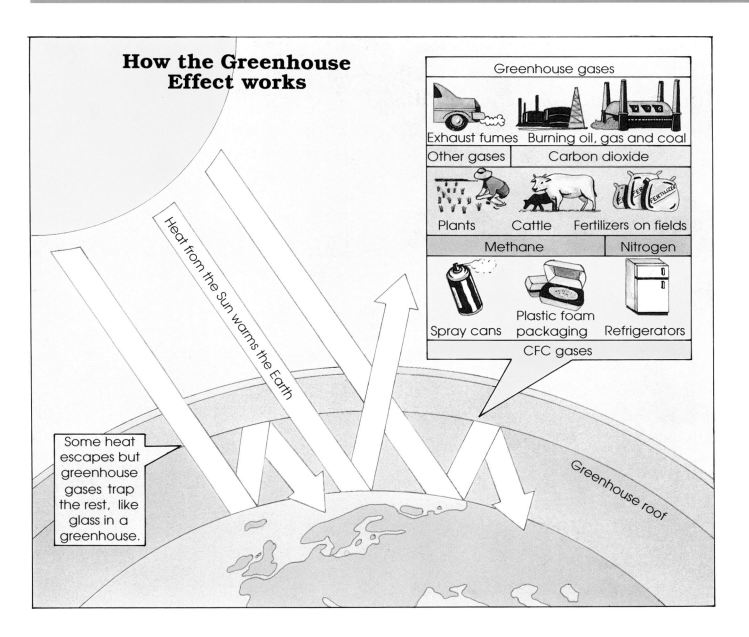

Today we are putting more carbon dioxide into the air. This happens because we are burning more fossil fuels to provide energy. More carbon dioxide in the atmosphere means less heat can escape to space. So the Earth is getting warmer. This is called the 'greenhouse effect'.

◀ *Grazing cattle produce methane gas. This is a greenhouse gas.*

What happens to the smoke that comes out of power stations and factory chimneys? It goes into the air. But the air contains some moisture, which absorbs chemicals from the smoke. This pollutes the air. If it is windy, the polluted moisture is blown hundreds of kilometres. Then it falls to the ground as rain or snow. It is called **acid rain** because the smoke has made the water droplets acid.

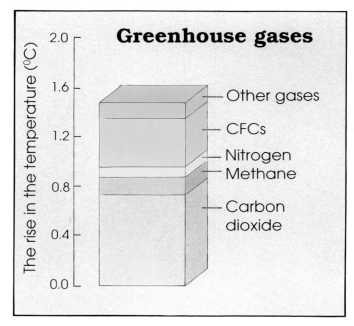

▲ *This diagram shows how greenhouse gases could all warm up the Earth.*

◀ *When lakes are damaged by acid rain, the fish die and there are none to catch.*

Acid rain causes many problems. It can kill the plants and animals living in lakes and streams. Many lakes in **Scandinavia** and North America are known as 'dead lakes'. Nothing lives in them because acid rain has made the water too acid.

Acid rain damages trees. In Czechoslovakia, whole forests have been killed by acid rain.

Acid rain also damages buildings. It eats away the stonework. Many statues on old cathedrals have been damaged by acid rain.

Acid rain is killing trees in this forest in the USA. ▼

▼ *This is the Parthenon in Athens, Greece. It has stood for over 2,000 years. Now acid rain is eating away the stonework.*

▲ *Car exhaust fumes pollute the air that we breathe.*

Trucks, buses, cars and tractors are called vehicles. There are over 550 million vehicles in the world. That is enough to go around the Earth forty times. Each vehicle has an engine which needs fuel to make it work. The fumes from its exhaust cause extra pollution.

▲ *All the people in cars in this picture could be carried in a few large buses. This would save fuel and pollution.*

Vehicles carry people and goods from one place to another. We cannot do without them. But exhaust fumes can hurt our eyes, give us a headache and make it hard to breathe. In big cities the Sun turns the fumes into an unpleasant and unhealthy haze, or **smog**, on hot, calm summer days.

A lot of fuel that is burnt is wasted. But we could easily save some of this fuel. We could use buses and trains more often. We could turn off lights that are not needed and turn down the heating slightly in our homes. Better **insulation** in our buildings would keep in more heat. Using less fuel means less air pollution.

It is possible to make energy from the power of the wind, waves and waterfalls, and from the heat of the Sun. All these cause little or no pollution.

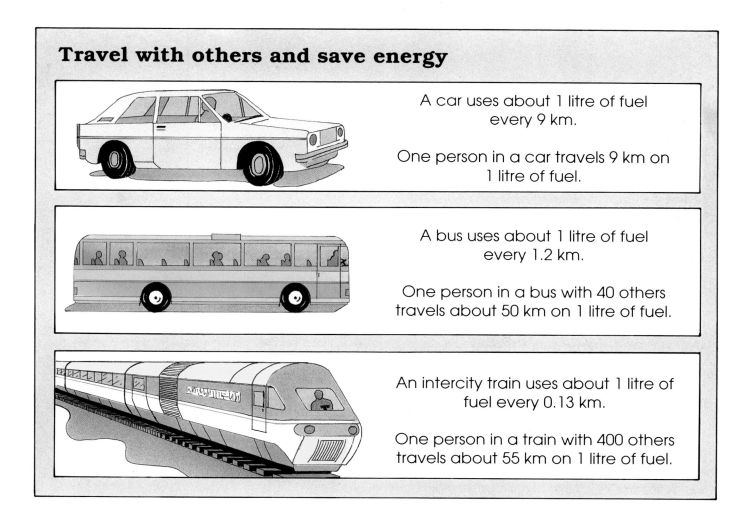

Travel with others and save energy

A car uses about 1 litre of fuel every 9 km.

One person in a car travels 9 km on 1 litre of fuel.

A bus uses about 1 litre of fuel every 1.2 km.

One person in a bus with 40 others travels about 50 km on 1 litre of fuel.

An intercity train uses about 1 litre of fuel every 0.13 km.

One person in a train with 400 others travels about 55 km on 1 litre of fuel.

▲ *Many Brazilian cars run on alcohol instead of petrol. Cars using this fuel cause little pollution.*

The fumes from cars can be cleaned before they reach the air. Many modern cars have a **catalytic converter** fitted into the exhaust pipe. It turns most of the poisonous fumes into carbon dioxide, nitrogen and water.

Factories can have special filters in the chimneys, to stop dangerous dirt and chemicals from going into the air.

Trees use carbon dioxide when they grow. If there were more trees there would be less carbon dioxide in the air. But many of our huge forests are being chopped down.

▲ *Every year huge areas of forest are cut down.*

▲ *Here, wood is burnt in big clay ovens to make charcoal. This puts more carbon dioxide into the air.*

Most countries of the world belong to the **United Nations**. **Politicians** from each country meet to try and solve world problems, such as air pollution. Scientists collect information about the atmosphere. This helps the politicians to decide what to do.

▼ *World leaders work together at the United Nations.*

◀ *These people belong to Greenpeace. Their banner, in central London, drew attention to the acid rain problem.*

There are other international organizations working to protect the environment. They include the World Conservation Union, Friends of the Earth and the World Wide Fund for Nature (WWF).

Cleaner cars

Drivers in busy cities like Tokyo and Los Angeles have some of the cleanest cars. That is because these cities have very strict pollution laws.

Car exhausts are checked for the amount of pollution they cause. A car is taken off the road if it fails the test.

▲ *In 1989, Queen Elizabeth II announced that all her cars would run on unleaded petrol.*

Lead in petrol

Lead used to be added to petrol in very small amounts. It helped car engines to run better. But lead is also very poisonous and exhaust fumes can make people ill.

Now more and more drivers are using unleaded petrol.

Drivers should use unleaded petrol because:

- It causes less pollution.

- It is cheaper.

- It reduces the amount of poisonous lead in the air.

◀ *A driver fills up with unleaded petrol.*

How to reduce pollution

To stop so much pollution coming from their exhausts, drivers can:

- Service their car regularly. An engine that is working well produces less pollution.

- Drive no faster than 90 kph.

- Fit a catalytic converter in their car exhaust and use unleaded petrol.

- Use public transport, cycle or walk whenever possible.

Chemicals in the air

Ozone is a gas found in the air. A narrow layer of ozone surrounds the Earth about 20 km above the surface. The ozone layer is being damaged by air pollution. It sometimes disappears over the South Pole. The ozone layer also gets thinner over the **Arctic** regions in spring.

▼ *In this satellite photograph, the grey area over Antarctica (centre) shows a hole in the ozone layer. Fortunately, this hole does not last all year.*

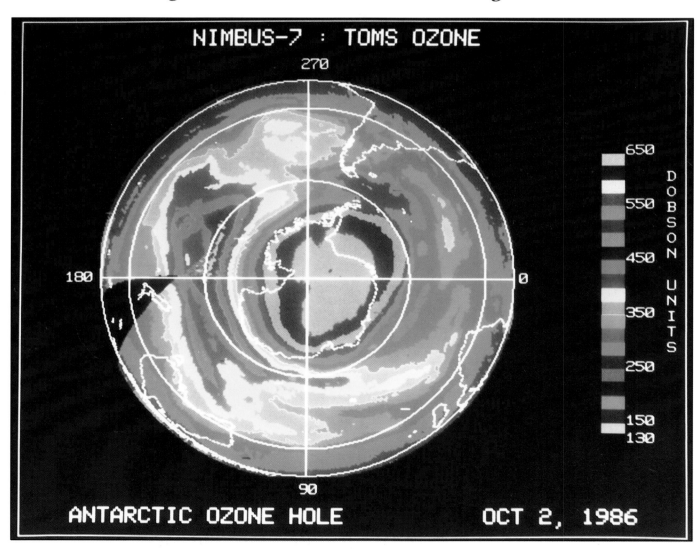

▲ *It would be dangerous to sunbathe if the hole in the ozone layer were overhead.*

The ozone layer is important because it absorbs some of the Sun's heat and keeps out some of the Sun's harmful ultraviolet rays. These can cause skin cancer and eye problems. The rays also harm tiny creatures in the sea, called plankton, which are food for many sea creatures.

With less ozone, the Earth will become warmer.

The ozone layer is damaged by chemicals called chlorofluorocarbons (CFCs). They are used in some spray cans and in the cooling systems of refrigerators. They are also used to make plastic foam and to clean the **microchips** used in computers.

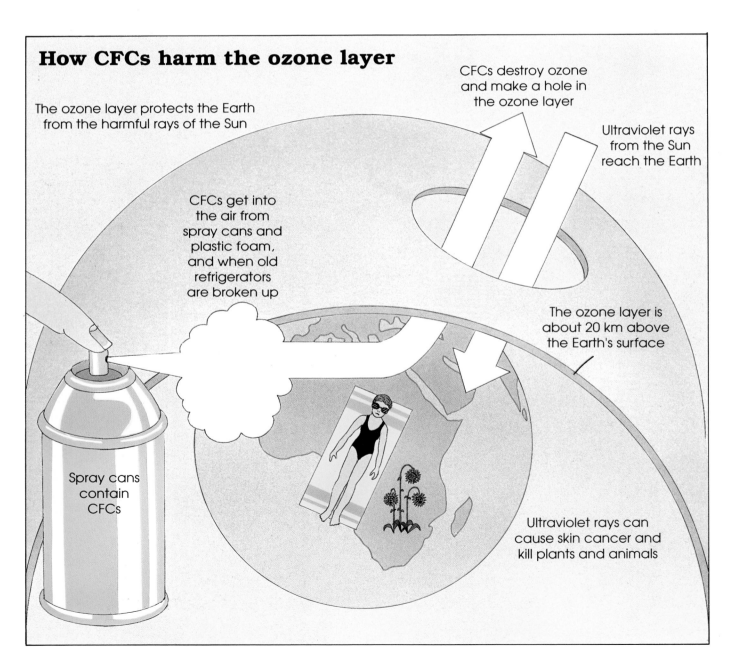

How CFCs harm the ozone layer

The ozone layer protects the Earth from the harmful rays of the Sun

CFCs destroy ozone and make a hole in the ozone layer

Ultraviolet rays from the Sun reach the Earth

CFCs get into the air from spray cans and plastic foam, and when old refrigerators are broken up

The ozone layer is about 20 km above the Earth's surface

Spray cans contain CFCs

Ultraviolet rays can cause skin cancer and kill plants and animals

◀ *Refrigerators contain CFCs. When do you think the CFCs escape into the atmosphere?*

When scientists discovered that CFCs were harmful, they looked for chemicals that would not damage the ozone layer. Today, most spray cans are 'ozone friendly'.

Big computer firms, like IBM, now clean microchips with products that do not contain CFCs. Some of the bigger hamburger chains, such as McDonalds, have stopped using CFCs in their plastic foam packaging.

▲ *If you need to buy a spray can, look for one marked 'ozone friendly'.*

Most of the countries that use CFCs have agreed to stop using them. By the end of the century, CFCs may not be used at all. However, it will still take about 50 years for the hole in the ozone layer to recover.

When you use washing-up liquid, you squirt chemicals into the water. Afterwards you pour the dirty water down the sink.

Washing-up liquid is fairly harmless. But other chemicals are much more dangerous and cannot just be thrown down the sink, into the drain or put on a rubbish dump.

The world's worst chemical accident happened in 1984 at Bhopal in India. Poisonous fumes killed over 2,000 people and many farm animals. ▶

Some chemicals have to be destroyed in a very hot fire to make them safe. Special ships have been built to burn these chemicals at sea, where pollution controls are not so strict.

Environmental groups say this is wrong. Sixty-five countries agree with them and are banning it from 1994.

▼ *This special ship is burning unwanted dangerous chemicals at sea.*

▲ *Greenpeace is against burning dangerous chemicals at sea, as their banner shows.*

We know how dangerous it can be to dirty the air with smoke, fumes and chemicals. Even burning garden and household rubbish can put dangerous chemicals into the air. Scientists are always trying to find new ways to prevent pollution. Politicians are passing more laws to stop people and industries from making the air dirty.

We spray
chemicals on
wood to kill
woodworm
and other pests.
But the strong
chemicals can
also kill bats,
which do no
harm at all. ▶

◀ Paints, old engine oil and
other chemicals should
be taken to a special
waste collection site.

Nuclear energy

Nuclear power stations use a metal called uranium for fuel. It is not burnt like oil or coal. The **atoms** that make up uranium can be split to give off a huge amount of energy. This can be used to heat water, make steam, and so drive the **generators** which make electricity.

▼ *All energy sources have good and bad points.*

Sources of energy	Bad points	Good points
Coal/oil/gas power	Causes acid rain. Adds to the greenhouse effect. Fossil fuels will run out.	Cheap electricity. Safe. Possible to control pollution (though expensive).
Windmills and wind power	Noisy. Many windmills so large areas of land needed. Can spoil countryside.	No air pollution. Fuel (wind) is free and will not run out.
Water power	Large areas of land have to be flooded. This can harm wildlife and spoil countryside.	No air pollution. Fuel (water) is free. Creates new surroundings for water creatures.
Wave power	Needs large area of sea. Equipment could be a danger to shipping.	No air pollution. Fuel (waves) is free and will not run out.
Nuclear power	Produces radioactive waste. Nuclear accidents dangerous to animal and plant life.	No air pollution unless radioactivity leaks. Fuel (uranium) will not run out.
Solar energy	Solar panels take up large area of land.	No air pollution. Fuel (heat of the Sun) is free and will not run out.

Nuclear power – yes or no?

Some people think we should build more nuclear power stations. They say:

- Unlike fossil fuels, uranium will never run out.
- Nuclear power stations do not pollute the air with smoke and fumes.
- Nuclear power stations do not put carbon dioxide into the air, so they do not make the Earth warmer.

Other people say we should stop using nuclear power because:

- Nuclear power puts more **radioactivity** into the air, and radioactivity can cause cancer.
- Nuclear accidents are very dangerous.
- The used uranium is radioactive and needs to be stored safely for thousands of years.

▼ *These people do not want dangerous radioactive waste buried near their homes.*

▼ *In 1986 a serious explosion happened at the nuclear power station in Chernobyl, USSR.*

What else pollutes the air?

The air inside our homes can be very easily polluted, especially when all the doors and windows are closed.

People who smoke make the air unpleasant for other people. Coal, wood or gas fires can send fumes into a room if the chimney is partly blocked.

Chemicals, such as spray polishes, cleaners and glue, can pollute the air in our homes. Always use them carefully and store them safely with the lid on.

▼ *Chemicals kept at home can pollute the air we breathe.*

◄ *Some jet aircraft are very noisy.*

Loud noise is a form of pollution. It can damage our hearing. Many countries have laws to reduce noise.

▼ *Many people dislike cigarette smoke, especially at mealtimes.*

▲ *Smog in Los Angeles, USA, used to be common. Now the air is cleaner because pollution from cars has been reduced.*

The atmosphere around the Earth makes it possible for plants and animals to live here. This includes you! It stops harmful rays from the Sun reaching the surface of the Earth. It affects our climate and weather.

But we are damaging the atmosphere. We put smoke, fumes and dangerous chemicals into the air. This harms plants and animals. The climate and weather are changing and the ozone layer is being destroyed.

But remember, if we damage the air we damage ourselves. We must stop making the air dirty.
We can all help to stop air pollution. Will you help too?

▼ *We all enjoy clean, fresh air. Let's not spoil it.*

What you can do

At present, most of our energy comes from fossil fuels which cause pollution. These fuels will run out one day. We should all try to save energy. This will reduce air pollution and make the fossil fuels last longer.

You can help in the following ways:

- Put on extra clothes when it is cold, instead of turning up the heating. This will mean that less fuel is burnt.

- Turn off the lights when they are not needed, so that power stations do not have to produce so much electricity.

- Ask the drivers in your family to drive cars more slowly so that the engine produces less pollution.

- Use buses and trains instead of cars – they can carry far more people in one trip. Better still, walk or cycle whenever you can.

- Tell other people what you have learnt about air pollution, so that they can help too. You might like to write a letter to your local paper.

Can you think of any other ways?

The greenhouse effect

If we continue to pollute the atmosphere, the greenhouse effect will worsen. With a group of friends, try to imagine how your life might be different if:
- summers were hotter, drier and longer
- winters had much colder spells
- there was more rainfall
- there were hurricane-force winds at least once a year.

Glossary

Absorbs   Takes up something (for example, a sponge absorbs water).

Acid rain   Rain, snow and mist that has become acid from pollution.

Arctic   The very cold region of the world near the North Pole.

Atmosphere   The mixture of gases that surrounds the Earth.

Atoms   Very tiny particles that group together to form all things. A thousand million atoms are only as large as this full stop.

Carbon dioxide   A gas. We produce it when we breathe out and when we burn fuels like coal, oil and gas.

Catalytic converter   A device fitted in car exhausts to reduce pollution.

Environmental group   A group of people that work together to try to protect our surroundings.

Fossil fuels   Sources of fuel such as coal, oil and natural gas, which have been formed over thousands of years from the remains of dead animals and plants.

Fumes   Poisonous gases.

Gases   Substances that have no shape and are not liquid or solid. Air is a mixture of gases.

Generator   A machine that makes electricity.

Insulation   Materials that stop heat escaping.

Microchips   The very small electrical circuits used in computers.

Politicians The people in charge of making laws.

Pollution   Anything that damages our surroundings. Pollution can spoil land, water and air.

Power stations   Factories that produce electricity.

Radioactivity A kind of energy which can damage living things.

Scandinavia The northern European countries of Norway, Sweden, Iceland, Finland and Denmark.

Smog A mixture of fog, fumes and smoke.

United Nations The international organization which brings all the countries of the world together to try and solve world problems.

Finding out more

Books to read

Acid Rain by Tony Hare (Franklin Watts, 1990)
Acid Rain by Stephen Sterling (Wayland, 1991)
Acid Rain by John Baines (wallchart and notes – Pictorial Charts Educational Trust, 1989)
Alternative Energy – five-book series (Wayland, 1990)
The Blue Peter Green Book by Lewis Bronze, Nick Heathcote and Peter Brown (BBC Books, 1990)
The Green Detective up the Chimney by John Baines (Wayland, 1991)
Lichens and Air Pollution – colour wallchart with pictures of common lichens, describing how they show the level of air pollution in a given area (BP Educational Service)
The Young Green Consumer Guide by John Elkington and Julia Hailes (Gollancz, 1990)

Useful addresses

Acid Rain Foundation
1630 Blackhawk Hills
St Paul MN 55122, USA

Acid Rain Information Centre
Dept of Environment and
    Geography
Manchester Polytechnic
Chester Street
Manchester M1 5GD

Campaign for Lead-Free Air
    (CLEAR)
3 Endsleigh Street
London WC1H 0DD

Council for Environmental
    Education
University of Reading
London Road
Reading RG1 5AQ

Friends of the Earth (UK)
26-28 Underwood Street
London N1 7JQ

Friends of the Earth (Australia)
National Liaison Office
366 Smith Street
Collingwood
Victoria 3065

Friends of the Earth (Canada)
Suite 53
54 Queen Street
Ottawa KP5CS

Friends of the Earth (New
    Zealand)
Negal House
Courthouse Lane
PO Box 39/065
Auckland West

Greenpeace (UK)
30-31 Islilngton Green
London N1 8XE

The Kids' EarthWorks Group
1400 Shattuck Avenue, 25
Berkeley CA 94709, USA

National Society for Clean Air
136 North Street
Brighton
East Sussex BN1 1RG

World Wide Fund for Nature (UK)
Panda House
Weyside Park
Catteshall Lane
Godalming
Surrey GU7 1XR

Index

Picture acknowledgements
Associated Press 28; Bruce Coleman Ltd *cover* (Stephen J Krasemann), 6 (Hans Reinhard), 10 below (Hans-Peter Merten), 11 (B & C Alexander), 12 (D Houston), 19 (Cliff Hollenbeck), 20 above, 25 below (L C Marigo); Greenpeace 35, 36; Hutchison Library 14 above (P Edward Parker), 22 (R Ian Lloyd), 24 (Caran McCarthy), 31 (Bernard Regent); ICCE 29 (Mike Hoggett); Oxford Scientific Films 25 above (Sean Morris), 37 below (Doug Wechsler), 37 above (Michael Leach); Photri 7, 10 above; Rex Features Ltd 33 above, 34 (Haley), 39 (Bradwell); Paul Seheult 40; Frank Spooner 27 (Julian Parker); Topham 13, 14 below, 15, 20 below, 26, 39 left, 41 below; ZEFA 4 (Mueller), 5 (C Voigt), 21, 30, 33 below, 41 above (G Kalt), 42 (J O'Rourke), 43 (K Kerth). The illustrations are by Marilyn Clay.